UNDER THE MICROSCOPE

SCIENCE TOOLS

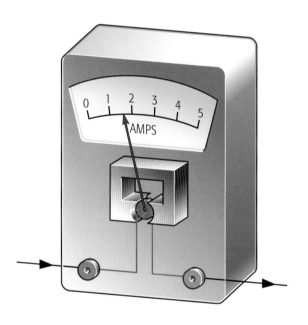

4 ELECTRICAL MEASUREMENT

John O.E. Clark

GROLIER
EDUCATIONAL

About this set

SCIENCE TOOLS deals with the instruments and methods that scientists use to measure and record their observations. Theoretical scientists apply their minds to explaining a whole range of natural phenomena. Often the only way of testing these theories is through practical scientific experiment and measurement—which are achieved using a wide selection of scientific tools. To explain the principles and practice of scientific measurement, the nine volumes in this set are organized as follows:

Volume 1—Length and Distance; Volume 2—Measuring Time; Volume 3—Force and Pressure; Volume 4—Electrical Measurement; Volume 5—Using Visible Light; Volume 6—Using Invisible Light; Volume 7—Using Sound; Volume 8—Scientific Analysis; Volume 9—Scientific Classification.

The topics within each volume are presented as self-contained sections, so that your knowledge of the subject increases in logical stages. Each section is illustrated with color photographs, and there are diagrams to explain the workings of the science tools being described. Many sections also contain short biographies of the scientists who discovered the principles that the tools employ.

Pages at the end of each book include a glossary that gives the meanings of scientific terms used, a list of other sources of reference (books and websites), and an index to all the volumes in the set. There are cross-references within volumes and from volume to volume at the bottom of the pages to link topics for a fuller understanding.

Published 2003 by Grolier Educational, Danbury, CT 06816

This edition published exclusively for the school and library market

Planned and produced by Andromeda Oxford Limited,
11-13 The Vineyard,
Abingdon, Oxon OX14 3PX

Copyright © Andromeda Oxford Limited

Project Director Graham Bateman
Editors John Woodruff, Shaun Barrington
Editorial assistant Marian Dreier
Picture manager Claire Turner
Production Clive Sparling

Design and origination by Gecko

Printed in Hong Kong

Library of Congress Cataloging-in-Publication Data

Clark, John Owens Edward.
  Under the microscope : science tools / John O.E. Clark.
    p. cm.
Summary: Describes the fundamental units and measuring devices that scientists use to bring systematic order to the world around them.
Contents: v. 1. Length and distance -- v. 2. Measuring time -- v. 3. Force and pressure -- v. 4. Electrical measurement -- v. 5. Using visible light -- v. 6. Using invisible light -- v. 7. Using sound -- v. 8. Scientific analysis -- v. 9. Scientific classification.
  ISBN 0-7172-5628-6 (set : alk. paper) -- ISBN 0-7172-5629-4 (v. 1 : alk. paper) -- ISBN 0-7172-5630-8 (v. 2 : alk. paper) -- ISBN 0-7172-5631-6 (v. 3 : alk. paper) -- ISBN 0-7172-5632-4 (v. 4 : alk. paper) -- ISBN 0-7172-5633-2 (v. 5 : alk. paper) -- ISBN 0-7172-5634-0 (v. 6 : alk. paper) -- ISBN 0-7172-5635-9 (v. 7 : alk. paper) -- ISBN 0-7172-5636-7 (v. 8 : alk. paper) -- ISBN 0-7172-5637-5 (v. 9 : alk. paper)
  1. Weights and measures--Juvenile literature. 2. Measuring instruments--Juvenile literature. 3. Scientific apparatus and instruments--Juvenile literature. [1. Weights and measures. 2. Measuring instruments. 3. Scientific apparatus and instruments.] I. Title: Science tools. II. Title.
  QC90.6 .C57 2002
  530.8--dc21
                    2002002598

About this volume

Volume 4 of *Science Tools* is about measuring quantities involved with electricity. In static—stationary—electricity scientists use various instruments to measure electric charge. The quantities measured in moving electricity, which involves the flow of electricity along wires, include current, voltage, and resistance. They are all measured using some form of meter—ammeter, voltmeter, or ohmmeter. Another electricity meter, found in almost every home, records the amount of electric power used. It is called a wattmeter. Another type of meter measures magnetic fields that interact with electricity. These meters are found in the metal detectors used in airports and to hunt for buried treasure.

Contents

Main units of measurement

Scientists spend much of their time looking at things and making measurements. These observations allow them to develop theories, from which they can sometimes formulate laws. For example, by observing objects as they fell to the ground, the English scientist Isaac Newton developed the law of gravity.

To make measurements, scientists use various kinds of apparatus, which we are calling "science tools." They also need a system of units in which to measure things. Sometimes the units are the same as those we use every day. For instance, they measure time using hours, minutes, and seconds—the same units we use to time a race or bake a cake. More often, though, scientists use special units rather than everyday ones. That is so that all scientists throughout the world can employ exactly the same units. (When they don't, the results can be very costly. Confusion over units once made NASA scientists lose all contact with a space probe to Mars.) A meter is the same length everywhere. But everyday units sometimes vary from country to country. A gallon in the United States, for example, is not the same as the gallon people use in Great Britain (a U.S. gallon is about one-fifth smaller than a UK gallon).

On these two pages, which for convenience are repeated in each volume of *Science Tools*, are set out the main scientific units and some of their everyday equivalents. The first and in some ways most important group are the SI units (SI stands for Système International, or International System). There are seven base units, plus two for measuring angles (Table 1). Then there are 18 other derived SI units that have special names. Table 2 lists the 11 commonest ones, all named after famous scientists. The 18 derived units are defined in terms of the 9 base units. For example, the unit of force (the newton) can be defined in terms of mass and acceleration (which itself is measured in units of distance and time).

▼ Table 1. **Base units** of the SI system

QUANTITY	NAME	SYMBOL
length	meter	m
mass	kilogram	kg
time	second	s
electric current	ampere	A
temperature	kelvin	K
luminous intensity	candela	cd
amount of substance	mole	mol
plane angle	radian	rad
solid angle	steradian	sr

▼ Table 2. **Derived SI units** with special names

QUANTITY	NAME	SYMBOL
energy	joule	J
force	newton	N
frequency	hertz	Hz
pressure	pascal	Pa
power	watt	W
electric charge	coulomb	C
potential difference	volt	V
resistance	ohm	Ω
capacitance	farad	F
conductance	siemens	S
inductance	henry	H

▼ **Table 3. Metric prefixes** for multiples and submultiples

Prefix	Symbol	Multiple
deka-	da	ten ($\times 10$)
hecto-	h	hundred ($\times 10^2$)
kilo-	k	thousand ($\times 10^3$)
mega-	M	million ($\times 10^6$)
giga-	G	billion ($\times 10^9$)

Prefix	Symbol	Submultiple
deci-	d	tenth ($\times 10^{-1}$)
centi-	c	hundredth ($\times 10^{-2}$)
milli-	m	thousandth ($\times 10^{-3}$)
micro-	μ	millionth ($\times 10^{-6}$)
nano-	n	billionth ($\times 10^{-9}$)

Scientists often want to measure a quantity that is much smaller or much bigger than the appropriate unit. A meter is not much good for expressing the thickness of a human hair or the distance to the Moon. So there are a number of prefixes that can be tacked onto the beginning of the unit's name. The prefix milli-, for example, stands for one-thousandth. Therefore a millimeter is one-thousandth of a meter. Kilo- stands for one thousand times, so a kilometer is 1,000 meters. The commonest prefixes are listed in Table 3.

Table 4 shows you how to convert from everyday units (known as customary units) into metric units, for example from inches to centimeters or miles to kilometers. Sometimes you may want to convert the other way, from metric to customary. To do this, divide by the factor in Table 4 (not multiply). So, to convert from inches to centimeters, *multiply* by 2.54. To convert from centimeters to inches, *divide* by 2.54. More detailed listings of different types of units and their conversions are given on pages 6–7 of each volume. You do not have to remember all the names: They are described or defined as you need to know them throughout *Science Tools*.

To convert from	To	Multiply by
inches (in.)	centimeters (cm)	2.54
feet (ft)	centimeters (cm)	30.5
feet (ft)	meters (m)	0.305
yards (yd)	meters (m)	0.914
miles (mi)	kilometers (km)	1.61
square inches (sq in.)	square centimeters (sq cm)	6.45
square feet (sq ft)	square meters (sq m)	0.0930
square yards (sq yd)	square meters (sq m)	0.836
acres (A)	hectares (ha)	0.405
square miles (sq mi)	hectares (ha)	259
square miles (sq mi)	square kilometers (sq km)	2.59
cubic inches (cu in.)	cubic centimeters (cc)	16.4
cubic feet (cu ft)	cubic meters (cu m)	0.0283
cubic yards (cu yd)	cubic meters (cu m)	0.765
gills (gi)	cubic centimeters (cc)	118
pints (pt)	liters (l)	0.473
quarts (qt)	liters (l)	0.946
gallons (gal)	liters (l)	3.79
drams (dr)	grams (g)	1.77
ounces (oz)	grams (g)	28.3
pounds (lb)	kilograms (kg)	0.454
hundredweights (cwt)	kilograms (kg)	45.4
tons (short)	tonnes (t)	0.907

▶ **Table 4. Conversion** to metric units

Units used in electricity

The science of electricity is the study of electric charges, currents, voltages, and resistances. Each of these has its own unit, and each unit is named after a famous scientist: two Frenchmen, Charles de Coulomb and André Ampère, an Italian, Alessandro Volta, and a German, Georg Ohm.

Electric power plants supply electricity to homes, offices, and factories. The electricity travels along wires made of metal, usually copper. Silver is the only metal that carries electricity better than copper does, but it is far too expensive to use silver wire for thousands of kilometers of power lines. A material through which electricity can be made to flow is called a conductor. Copper, like many other metals, is a good conductor because it has lots of what are called free electrons. (An electron is a tiny subatomic particle that has a negative electric charge.) An electric current flows when these free electrons move along a wire made from a conductor.

A difference in voltage between the ends of a piece of wire makes the free electrons move along it. A battery can be used to create this voltage difference. The scientists who first investigated electricity decided that the current flows out of the positive terminal of the battery and returns to the negative terminal. They didn't know about electrons, and unfortunately they got their directions wrong.

We now know that electrons flow the other way, from negative to positive, but we still stick with the idea that current flows from positive to negative.

If a current that leaves a battery's positive terminal is going to get back to the negative terminal, there has to be an unbroken pathway between the two terminals. Such a pathway is called an electric circuit. The voltage of the battery "drives" the current around the circuit. The current will flow through anything else that is connected into the circuit, such as a bulb, as long as the circuit is complete. If there is a switch in the circuit, it has to be "on" for the current to flow. When the switch is set to the "off" position, the circuit is broken, and no current can flow.

► **The actual circuit** illustrated here shows a bulb connected to a battery by two wires. The battery is the source of voltage, and the bulb filament provides a resistance.

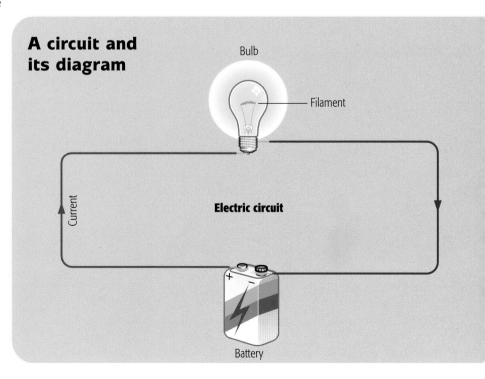

A circuit and its diagram

Bulb

Filament

Current

Electric circuit

Battery

Conductors and insulators

A material that electricity flows through easily is called a conductor. Copper and most other metals are good conductors. A material that does not allow electricity to pass through it, such as plastic or glass, is called a nonconductor or an insulator. The conducting power of some materials is in between these extremes, and they are termed poor conductors.

Because a poor conductor *resists* the flow of electricity, we say it has a resistance. Some dense metals such as tungsten have a high resistance. When a current is made to flow through tungsten, the metal gets hot. That is why it is used for making the filaments in light bulbs—the filaments get so hot that they give off white light.

Electrical units

Each electrical quantity has its own scientific unit. Electric charge, such as the charge on an electron, is measured in coulombs. The electron's charge is tiny,

equal to about one ten-million-million-millionth of a coulomb. During a thunderstorm a large thundercloud can store about 10 coulombs of electric charge. The coulomb (symbol C) was named for the French physicist Charles de Coulomb.

The unit of current, the ampere (symbol A), was named for another Frenchman, the physicist André-Marie Ampère. The ampere—usually shortened to "amp"—is quite a large unit, and small currents are generally measured in milliamperes (milliamps for short), one-thousandth of an amp (symbol mA). The appliances in your home use different amounts of current. An oven may need as much as 20 amps, and a water heater about 12 amps, so thick cables are used for the wires that carry the current to them. Other appliances use much less current. A vacuum cleaner, for example, uses 1 amp, while a television needs only about two-thirds of an amp.

Voltage is also sometimes called potential difference because it is the difference in electric

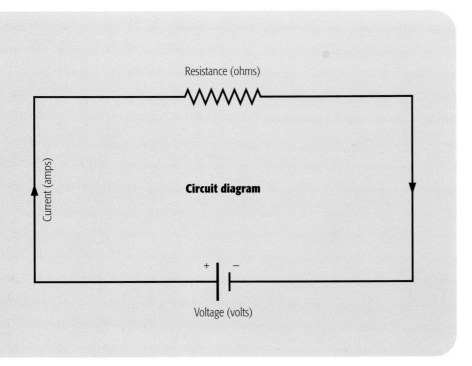

◄ **The equivalent circuit diagram** uses standard symbols to represent the battery and the bulb's resistance. The current flows from the positive terminal of the battery, around the circuit, and back to the battery's negative terminal.

Units used in electricity

▲ **A worker at an electricity substation** uses a long hook with an insulated handle to disconnect a high-voltage cable.

potential between two points in a circuit. It is measured in volts (symbol V), named for the Italian scientist Alessandro Volta, who made the first battery. Small voltages are measured in millivolts (mV, a thousandth of a volt) and large voltages in kilovolts (kV, a thousand volts) or megavolts (MV, a million volts). The standard voltage of the electricity supply to most American homes is 240 volts, though industrial plants often use much higher voltages.

The fourth common electrical unit is the ohm, the unit of resistance. Its symbol is Ω, the capital Greek letter omega. Again, larger units are formed by adding the usual metric prefixes: 1 kilohm (kΩ) = 1,000 ohms, and 1 megohm (MΩ) = 1,000,000 ohms. The resistance of the element in a one-bar electric heater, for example, is about 4 ohms, while the resistance of the filament in a 100-watt light bulb is one-tenth of an ohm. The ohm was named for the 19th-century German physicist Georg Ohm.

Other units

A component that stores electric charge is called a condenser or a capacitor, and its ability to store charge is termed its capacitance. Capacitance is measured in farads (symbol F), named for the British scientist Michael Faraday. The farad is a large unit, and most practical condensers have capacitances measured in millionths of a farad (microfarads, symbol μF) or even million-millionths of a farad (picofarads, symbol pF).

The tendency of a component in a circuit to produce a voltage because of nearby changes in current or magnetic field is called its inductance. It is measured in henrys (symbol H), named for the American physicist Joseph Henry.

The power consumed or produced by an electrical device is equal to the voltage multiplied by the current (volts × amps) and is measured in

► **Electricity is produced** by turbine generators. The heat from a burning fuel, such as coal or oil, heats water to produce steam. The steam spins turbines attached to electric generators. At full speed this generator at Colstrip, Montana, produces 350 megawatts (350 million watts) of electric power.

watts (symbol W). This unit of power was named for the Scottish engineer and inventor James Watt. Smaller and larger units are the milliwatt (mW, a thousandth of a watt), kilowatt (kW, a thousand watts), megawatt (MW, a million watts), and even gigawatt (GW, a billion watts). The electric power consumed by a domestic icebox, for example, is only about 40 watts, while a toaster or an electric iron uses 1,000 watts (1 kilowatt).

Derived units

Of all the electrical units described here, only the ampere—the unit of current—is a base unit in the SI system. All the others are derived units, defined in terms of one another. For example, a volt is defined as the potential difference between two points along a wire that produces 1 watt of power when a constant current of 1 amp flows between the points. In general, potential difference, or voltage, equals power divided by current (volts = watts ÷ amps).

An ohm is the resistance between two points along a wire when a constant potential difference of 1 volt between the two produces a current of 1 amp. In general, resistance equals voltage divided by current (ohms = volts ÷ amps).

And the watt itself is the power produced by a current of 1 amp flowing through a conductor that has a potential difference of 1 volt between its ends. In other words, watts equal volts times amps. Notice that the ampere, the SI base unit, appears in all of these expressions.

Measuring charge

There are two types of electric charge—positive charge and negative charge. Like most things in electricity, charge has to do with electrons. An object gets a positive charge when it loses one or more electrons, and it gets a negative charge when it gains one or more electrons.

▲ **A bad hair day** is guaranteed if you charge your hair with static electricity by touching a charged conductor.

One of the basic facts of electricity is that similar charges—two positive charges or two negative ones—repel each other and tend to move apart. But opposite charges—a positive charge and a negative one—attract each other and tend to move together. This is often summed up as "Like charges repel; unlike charges attract." The girl in the picture on the left has just touched a metal sphere that has a large electric charge on it. The charge has moved through her body to her hair. But because each hair has the same charge, the hairs all repel one another.

Charge detectors

The oldest and simplest science tool for detecting electric charge makes good use of "Like charges repel." Electric charges are invisible. The electroscope (illustrated on the right) got its name, which means an instrument for seeing charge, because it makes the presence of charge visible. An old-style gold-leaf electroscope (top right) is basically a glass jar on an insulating base. A metal rod passes down through a rubber stopper in the top of the jar. At the top of the rod is a brass disk, and two pieces of very thin gold foil (called leaves) are attached to the bottom of the rod.

Gold-leaf electroscope

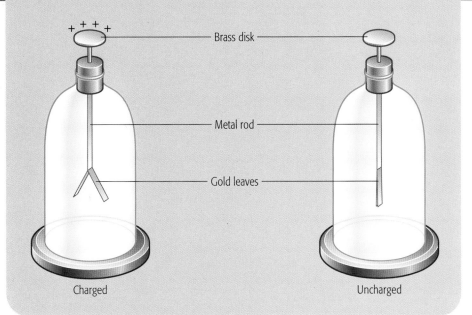

Brass disk

Metal rod

Gold leaves

Charged

Uncharged

◄ **With a simple gold-leaf electroscope** the presence of an electric charge makes the two pieces of gold foil (leaves) move apart— because similar charges repel each other. In the uncharged electroscope the two gold leaves hang down together.

Quartz-fiber electrometer

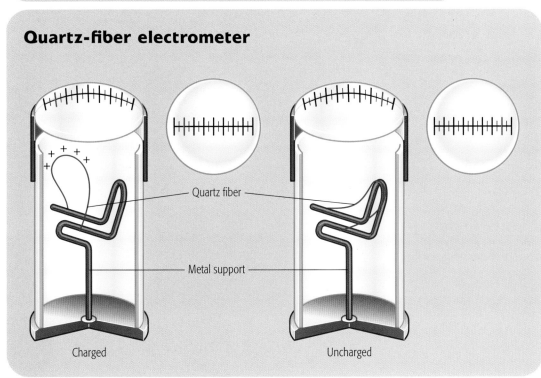

Quartz fiber

Metal support

Charged

Uncharged

◄ **A portable electrometer** is a miniature device carried by workers who risk being exposed to radiation. Before being used, the electrometer is charged, which makes the quartz fiber swing to the left. If radiation enters the device, the charge leaks away, and the fiber moves to the right. Its position as seen along the scale at the top indicates the radiation level.

You can give an electric charge to an object such as a rod made of glass or plastic by rubbing it with a piece of cloth or fur. When you move such a charged object close to an electroscope's disk, its charge is transferred to the disk. This charge travels down the metal rod to the gold leaves. Because both leaves get the same type of charge, they repel each other and move apart. If the charge is removed, for example, by touching the rod against the disk of the electroscope, the gold leaves lose their charge, they no longer repel each other, and they fall back to their normal position.

As well as detecting charge, an electroscope can be used to find out whether a charge is positive or negative. First, the instrument is given a charge of known type—positive, say. Then if another charged object—of unknown type—is moved close to the electroscope's disk, one of two things can happen. Either the gold leaves move even farther apart, which means they will have received more

FOR MORE ON MEASURING CHARGE SEE *MEASURING CURRENT* 4:14; *MEASURING VOLTAGE* 4:20; *X-RAYS AND GAMMA RAYS* 6:26

Measuring charge

charge, so the unknown charge was positive. Or the leaves collapse, which means the charge on the leaves will have been neutralized, and the unknown charge was negative.

Measuring charge

A different type of instrument, called an electrometer, not only detects electric charge but also measures it. The type in the lower illustration on page 11 is designed to be carried by people who risk being exposed to radiation, such as radiologists (who deal with medical x-rays) and workers in the nuclear industry. It is small enough to be mounted on the end of a tube and clipped inside a pocket, like a pen. In fact, it is often called a pen meter. Because the instrument measures doses of radiation, its proper name is a dosimeter.

The movable part of the electrometer is a thin fiber made of quartz, which is an insulating substance resembling glass. The quartz fiber is mounted on a metal support that is connected to

a voltage supply and charged to between 100 and 200 volts before use. The quartz fiber gets the same charge, and because like charges repel, it moves away from its support. Its position can be seen on the scale at the end of the instrument.

If high-energy radiation in the form of x-rays or gamma rays enters the meter's walls, it ionizes some of the molecules in the air, which allows some of the charge on the quartz fiber to leak away (to the walls). There is then less repulsion between the fiber and its support, so the two move closer together. The new position can be seen on the scale, which in this way gives a measure of the radiation exposure.

The name "electrometer" is also given to an electronic instrument for measuring voltages that doesn't draw much current itself. It is in effect a voltage amplifier. The device just described *is* a type of electrometer because what it measures is the voltage difference between the quartz fiber and its support.

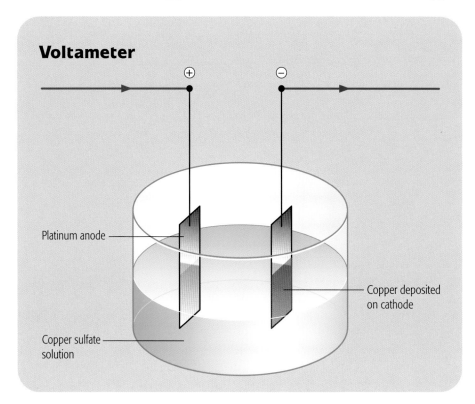

Voltameter

Platinum anode

Copper deposited on cathode

Copper sulfate solution

◄ **A voltameter** (not to be confused with a voltmeter, which measures voltages) measures charge by using electrolysis. Two platinum electrodes dip into a solution of copper sulfate. When an electric current flows between the electrodes, copper is deposited on the negative electrode (the cathode). The mass of copper that is deposited is a measure of the charge.

▲ **A lightning flash** is a huge electric spark. It jumps to the ground from the base of a cloud charged to several million volts.

Measuring charge by weighing

Electrolysis is a process in which an electric current travels through a solution carried by charged ions (atoms that have lost or gained some electrons). The current enters and leaves the solution, called the electrolyte, through metal plates called electrodes. It enters via the positive electrode (called the anode) and leaves via the negative electrode (the cathode).

If metal ions are present, a change takes place at the cathode. Say, for example, that the electrolyte is a solution of copper sulfate, which contains copper ions. At the cathode, positively charged copper ions pick up electrons, lose their charge, and become copper metal, which sticks to the cathode. The cathode gets electroplated with copper, and the mass of copper deposited depends on the amount of charge lost. The mass of copper is found by weighing the cathode before and after electrolysis.

An electrolytic cell used to measure charge in this way is called a voltameter (or sometimes a coulometer). It need not involve a metal being deposited. For example, a voltameter that has acidified water as the electrolyte produces hydrogen gas at the cathode. By measuring the volume of gas produced, it is possible to calculate its mass and work out the charge.

Measuring current

An electric current is a flow of electrons along a conductor. Some materials conduct better than others. The more electrons that can flow, the better the conductor, and the larger the current. Current is measured in amperes, and an instrument for measuring current is called an ammeter.

Various physical effects are caused by a flowing electric current. For example, a current can create a magnetic field around the wire it flows through, or it can heat the wire. The heating effect of an electric current finds practical use in electric light bulbs and heaters. Current-measuring instruments—collectively known as ammeters—also make use of such effects. An instrument for measuring very small currents, of thousandths of an amp (milliamps), is called a milliammeter.

◀ **A pocket calculator** uses only a small electric current. This one is powered by solar cells, which turn daylight into electricity.

► **A moving-coil ammeter** has a coil of wire pivoted between the pole pieces of a magnet. When current flows through the coil, the coil rotates, and a pointer attached to the coil indicates the current on a scale.

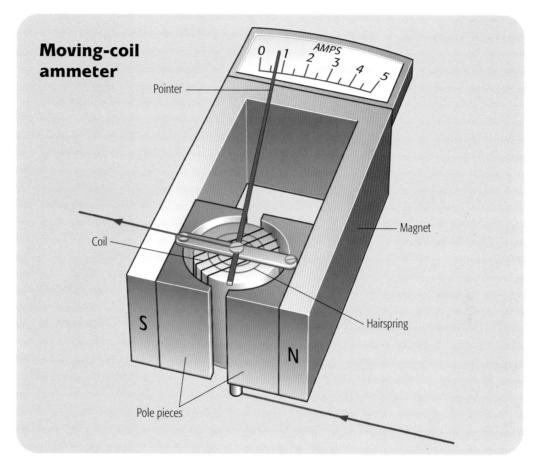

Moving-coil ammeter

Pointer

AMPS
0 1 2 3 4 5

Coil

Magnet

S

N

Hairspring

Pole pieces

Moving-coil ammeter

In the commonest type of ammeter, illustrated above, a coil of wire is made to move in a magnetic field. For this reason it is called—not surprisingly—a moving-coil ammeter.

The main part of the instrument—and the heaviest—is a magnet in the form of the letter U. Two curved pieces of iron, called pole pieces, are attached to the north and south poles of the magnet. A supporting iron cylinder carries a coil of thin insulated wire that is wound around it many times, end over end. A pointer is attached to the spindle of this cylinder, which is pivoted between the pole pieces.

When the meter is not in use, the pointer indicates zero on the scale. It is forced into this position by a fine hairspring that is coiled around the spindle.

When the ammeter is connected into an electric circuit, a current flows in the coil, which as a result produces its own magnetic field. The interaction between this field of the coil and the field of the magnet makes the coil try to rotate against the tension of the hairspring. The pointer is deflected and moves along the scale, and the distance moved depends on the strength of the current. Notice that the scale is linear—that is, it has equal divisions (the distance between 1 amp and 2 amps, for example, is the same as the distance between 4 amps and 5 amps). This makes it easy to read off the value of a current that falls between any two of the numbered points on the scale. Like all ammeters, the moving-coil ammeter has a very low electrical resistance, so that when it is connected into a circuit, its presence does not alter the current that is flowing around the circuit.

FOR MORE ON MEASURING CURRENT SEE *MEASURING CHARGE* **4**:*10*; *MEASURING VOLTAGE* **4**:*20*; *MEASURING ELECTRIC POWER* **4**:*32*

Measuring current

Hot-wire ammeter

As mentioned on page 14, another effect of an electric current flowing along a wire is to heat the wire. And when a metal wire gets hot, it expands and becomes longer. This expansion is what makes a hot-wire ammeter work. The construction of this type of ammeter is shown in the illustration below, which is of a demonstration instrument used in teaching laboratories.

The key part of the instrument is a length of high-resistance wire, shown in red in the illustration. When current flows through the ammeter, this wire gets hot and expands. A cord attached to a spring moves the wire down as it stretches. The same cord passes around a pulley located between the hot wire and the spring. There is a pointer attached to the center of the pulley.

As the hot wire stretches, the spring pulls the cord down, and that makes the pulley rotate slightly. The pointer then moves along the scale, which is calibrated (marked off) to indicate the current flowing through the instrument. Because the heating—and therefore the expansion—of the

wire is not directly proportional to the value of the current, the scale is not linear. That is, the divisions along the scale are not equal, and care has to be taken in reading off values of current that lie between those numbered on the scale. (If one quantity is directly proportional to another, then, for example, if one of them doubles, so does the other. If they are not directly proportional, then if one quantity doubles, the other will change by some other factor.)

Moving-iron ammeter

Like a moving-coil instrument, the moving-iron ammeter also makes use of the fact that an electric current flowing along a wire produces a magnetic field. This time the current flows around a large coil of wire. A laboratory version of the moving-iron ammeter is illustrated on the opposite page. "Real" versions have many more turns of wire in the coil than in this demonstration model.

Two pieces of iron are positioned near the bottom of the instrument. One of them is attached to the case, and the other is attached to the end of

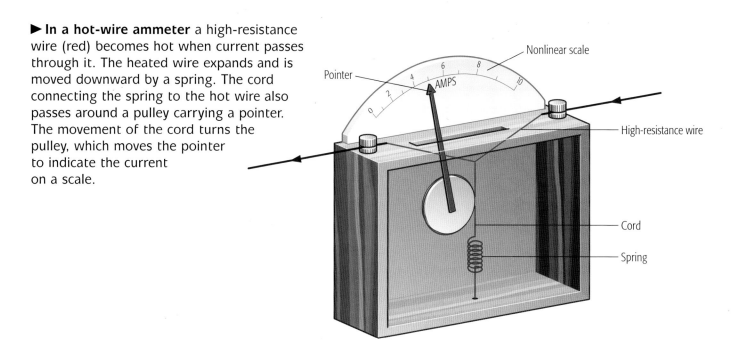

► **In a hot-wire ammeter** a high-resistance wire (red) becomes hot when current passes through it. The heated wire expands and is moved downward by a spring. The cord connecting the spring to the hot wire also passes around a pulley carrying a pointer. The movement of the cord turns the pulley, which moves the pointer to indicate the current on a scale.

Pointer

Nonlinear scale

AMPS

High-resistance wire

Cord

Spring

a long pointer, pivoted near the center of the back of the case. A fine hairspring coiled around the pointer's pivot keeps it pointing to the zero end of the scale.

The magnetic field produced by the current in the coil magnetizes both pieces of iron—in other words, they both become magnets. They get magnetized in the same sense, so that their north poles are located next to each other (as are their south poles). It is a basic law of physics that like magnetic poles repel each other. As a result, the newly created magnets repel and move apart. The fixed magnet cannot move, but the other one moves sideways, turning the pointer against the tension of the hairspring to indicate the current on a scale. Again (as in the hot-wire instrument) the magnetic effect is not directly proportional to current, so again the scale is nonlinear, with unequal divisions.

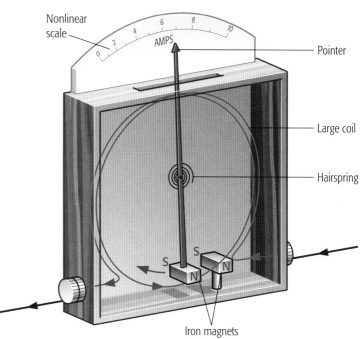

▲ **In a moving-iron ammeter** current flowing around a large coil of wire produces a magnetic field that magnetizes two pieces of iron, which then move apart.

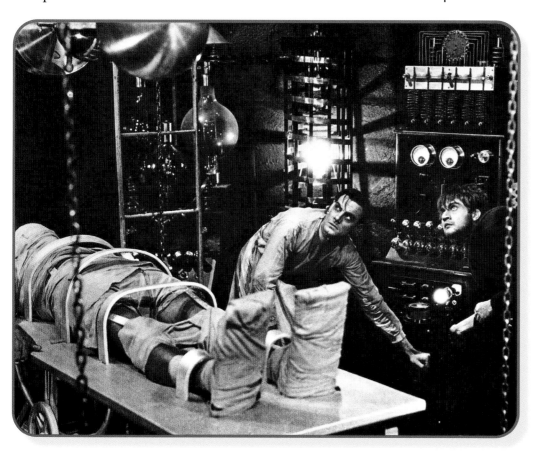

◄**The "scientists"** in this old *Frankenstein* movie seem too distracted by their experiment to look at the ammeters on their electrical apparatus!

Measuring current

Tangent galvanometer

A galvanometer is an instrument that detects very small electric currents. A tangent galvanometer does this but measures the strength of the current as well. It consists of a large coil of wire mounted vertically on a stand (see the illustration on the right). Inside the coil is a small horizontally positioned magnet on a pivot, with a long nonmagnetic pointer set at right angles to it.

Before use—that is, before the instrument is connected into an electric circuit—it is positioned so that the coil is in line with the Earth's magnetic field (the direction of the Earth's field varies from place to place). The small magnet will line up with the coil, so that the pointer is at right angles to the coil. The galvanometer is then connected into a circuit, and current flows around the large coil. This generates a magnetic field around the coil, and the field deflects the magnet and its pointer. The amount of deflection depends on the strength of the current, which is proportional to the tangent of the deflection angle (not the angle itself). This is not a problem if the scale is calibrated directly in milliamps.

Nowadays the tangent galvanometer is rarely used for measuring small currents, for which various electronic devices are available, but it is still used to measure the magnetizing force of the Earth's magnetic field.

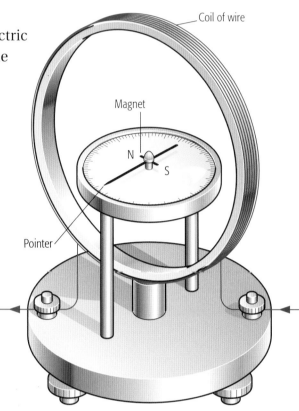

Coil of wire

Magnet

N
S

Pointer

▲ **In a tangent galvanometer** the current to be measured flows around a large coil of wire. The instrument is first aligned so that the coil is parallel to the Earth's magnetic field at the location where it is being used. When the current flows, a small magnetic needle at the center is deflected. It has a long pointer attached to it to make the deflection angle easier to read.

◀ **An electric current** will travel through a solution of chemicals, carried by ions, in the process called electrolysis. By measuring the current and carefully controlling it, workers can use electrolysis in electroplating, in which a metallic object is coated with a thin layer of another metal.

◄ **A worker lifts some panels** from an electroplating tank. The strength of the current controls the thickness of the coating that is applied to the panels to protect them against wear and corrosion.

André-Marie Ampère

Ampère was a French physicist and mathematician who founded the science of electromagnetism. He was born in Lyons in 1775, the son of a wealthy merchant who was later executed in the French Revolution. Ampère, who was self-taught, was an exceptionally good mathematician. After earning his living as a math teacher, from 1802 he held a series of professorships. In 1808 Napoleon made him inspector general of the universities, a position he held until his death in 1836.

He began his experiments on what he called "electrodynamics" (electromagnetism) in 1820, having heard of the discovery that a current passing through a wire could move the needle of a magnetic compass. Seven years later he announced what we now call Ampère's law, concerning the magnetic forces between wires that carry an electric current. The SI unit of current, which is named in his honor, is still defined in terms of the force between parallel wires carrying a current.

Measuring voltage

Voltage is the difference in electric potential between the ends of a conductor—it is the "force" that pushes electrons around an electric circuit. For this reason, voltage is also known as potential difference. An instrument for measuring voltage is called a voltmeter.

When the ammeters described on the last few pages are placed in an electric circuit, they are connected so that all the current flowing in the circuit passes through them—that is how they measure current. Ammeters have a very *low* resistance so that their presence in a circuit has little effect on the current flowing through the circuit. A voltmeter measures the voltage between two points in a circuit. To make sure that they do not affect the value of the voltage, voltmeters have a very *high* resistance. In fact, the simplest type of voltmeter is just an ammeter with a high resistance added to it.

Moving-coil voltmeter

The workings of a moving-coil ammeter are described in detail on page 15. A moving-coil voltmeter is a moving-coil ammeter that has a high resistance included in one of the internal leads to the moving coil. A voltmeter measuring up to 10 volts, for example, has a resistance of about 1,000 ohms. In practice, some measuring instruments are built as ammeters but come supplied with a resistance, called a shunt, that can be added so that they can be used as voltmeters. Care is needed because the instrument would be seriously damaged if it were wired into a circuit as a voltmeter without its protective resistance.

Measuring small voltages

Scientists use a potentiometer—the word means "potential measurer"—to measure small voltages in the laboratory. The apparatus consists of a resistance wire equipped with a sliding contact.

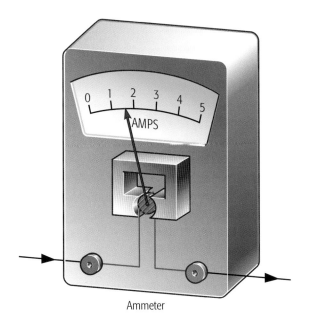

Ammeter

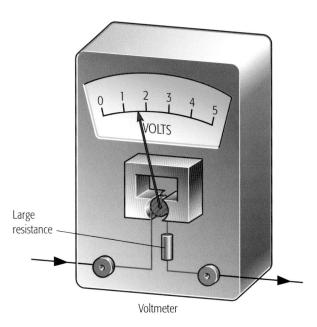

Voltmeter

▲ **A moving-coil voltmeter** is a sensitive ammeter (top) with a large resistance in the wire leading to the coil (bottom).

It is illustrated on page 22, where it is shown being used to measure the voltage of a small battery. The circuit includes a galvanometer, which is a very sensitive instrument that detects the presence of even the smallest electric current.

The output from an accumulator (the name for a rechargeable battery) is connected across the ends of the wire. The voltage to be measured, in this case the output of a small battery, is linked through the galvanometer to one end of the wire. The other lead from the small battery connects it to the contact that can slide along the wire. The sliding contact is moved until there is no deflection on the galvanometer. No deflection indicates that there is no current flowing in the small battery's circuit—its voltage has been exactly balanced by part of the accumulator's voltage.

The portion of the wire shown in red in the illustration divided by the total length of the wire gives the fraction of the accumulator's output that equals the small battery's voltage. For example, if the accumulator produces 6 volts, and the "red" portion of the wire is two-thirds of the total length, then the small battery's voltage is two-thirds of 6 volts, equal to 4 volts.

The oscilloscope

A voltmeter of the type described on the opposite page can be used for measuring steady voltages. But if we want to measure a voltage that varies

▲ **Voltmeters,** switches, and other measuring instruments surround the technicians in the control room of an electric power plant.

FOR MORE ON MEASURING VOLTAGE SEE *MEASURING CURRENT* 4:14; *MEASURING RESISTANCE* 4:26; *MEASURING ELECTRIC POWER* 4:32

Measuring voltage

with time—a voltage that is high one second, low the next—then we need a different instrument. An oscilloscope can measure and display such a varying voltage.

An oscilloscope is an electronic instrument built around a cathode-ray tube, similar to the tubes in radar displays, TVs, and computer monitors (see the illustration at the top of page 24). In a cathode-ray tube a device called an electron gun at the narrow end of the tube produces electrons by heating a filament and "fires" them at the screen at the opposite end of the tube. The inside of the screen is coated with a phosphor, which is a chemical that gives off light when struck by electrons. So wherever the electron beam strikes the screen, a spot of light is formed.

Inside the tube are two pairs of deflection plates mounted at right angles to each other. When a voltage is applied to a pair of plates, they deflect the electron beam passing between them. The horizontal plates make the electron beam move sideways, while the vertical plates move the beam up or down.

When the oscilloscope is used to examine a varying voltage, the horizontal deflection plates are

connected into a circuit (called a time base) that feeds in a repeating signal to make the spot of light on the screen trace a straight line from one side of the screen to the other. When it reaches one side, it flips back and draws the line over again. The varying voltage is fed to the vertical deflection plates, which make the spot of light move up and down in step with the voltage. But because the spot is also moving sideways, the combined effect is to trace a wavy line across the screen. You can see a "sawtooth" trace,

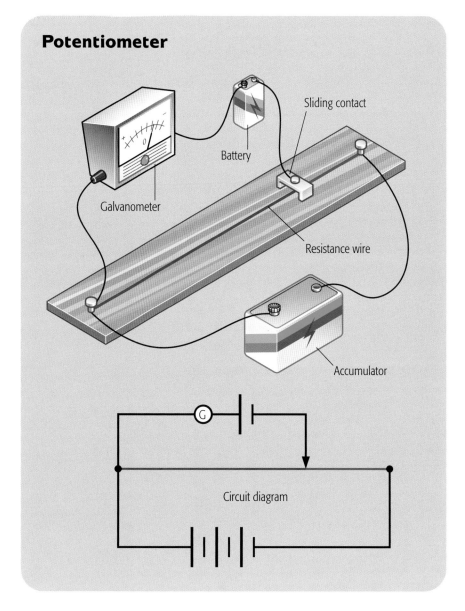

Potentiometer

Sliding contact

Battery

Galvanometer

Resistance wire

Accumulator

Circuit diagram

▶ **A potentiometer** measures the voltage of a battery by applying the output from an accumulator across a length of high-resistance wire and "tapping off" part of it via a sliding contact. A galvanometer indicates when the tapped-off voltage is the same as that of the battery being measured.

representing a voltage that rises and falls sharply, on the screen of the oscilloscope on the right of the picture below.

Measuring heartbeats

The oscilloscope is widely used for testing circuits and components in the electronics industry. But it is also important in medicine, as a science tool for examining the heart and the brain.

Your heart is a muscular organ that pumps blood around your body. When it beats, first the lower half contracts, squeezing the blood into the upper half. Then the upper half contracts, sending blood either to the lungs or out around the body. If you listen to someone's heartbeat, you can hear the two different stages of the pumping action as a "lub-dub" sound. To make the heart muscles contract, tiny electrical signals pass along nerves inside the heart. These voltages can be detected on the surface of the body by an instrument called an electrocardiograph (ECG), which displays a

Alessandro Volta

Volta's full name was Count Alessandro Giuseppe Antonia Volta. He was born in 1745 in Como, Italy. Unlike his male relatives, who nearly all became priests, Volta decided to study electricity. In 1774 he invented the electrophorus, a device for storing static electricity. But his best-known invention was the first electric battery, which consisted of a pile of alternating disks of copper and zinc separated by pieces of cloth soaked in salt solution. It became known as a voltaic pile. Volta died in 1827, some years before the unit of potential difference was named the volt to honor his achievement.

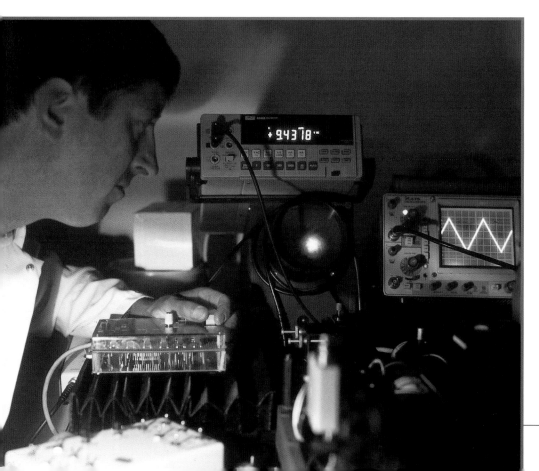

◄ **A scientist uses an oscilloscope** (on the far right) and a laser to check the voltage of some electronic equipment. The voltage is displayed on the digital voltmeter near the top, which is accurate to four decimal places.

Measuring voltage

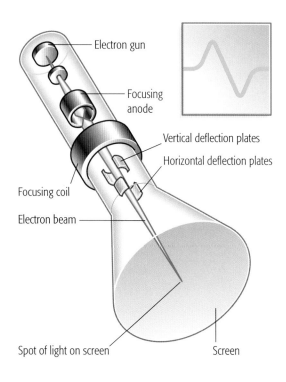

Electron gun

Focusing anode

Vertical deflection plates

Horizontal deflection plates

Focusing coil

Electron beam

Spot of light on screen

Screen

◄ **A cathode-ray tube** is the key part of an oscilloscope. Voltages applied to the deflection plates make a spot of light trace a graph on the screen.

trace of the voltages on an oscilloscope screen. It can also drive a chart recorder to record a trace as a wavy graph on a paper chart. These traces are called electrocardiograms, and they can tell physicians a lot about the heart's condition.

To use an ECG machine, a nurse or technician places disk electrodes on the patient's chest and perhaps other parts of the body. The electrodes are linked by wires to the ECG, which then records the heart's electrical activity. The whole procedure is painless and takes only a few minutes.

Measuring brain waves

The heart is not the only part of the body that is electrically active. Every second, whether you are awake or asleep, thousands of nerve messages

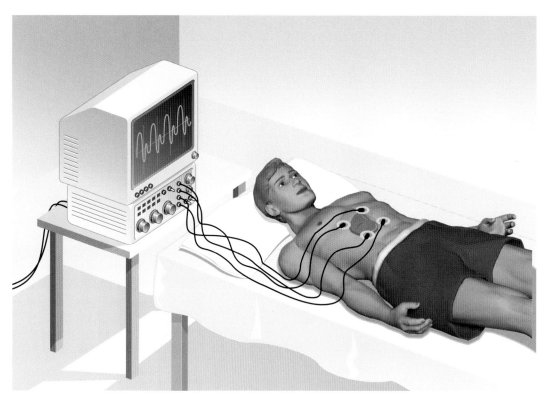

◄ **An electrocardio-graph (ECG)** picks up the electrical activity of the heart through electrodes placed on the person's chest. The resulting voltages are displayed on a computer screen or recorded by pens on a moving strip of paper.

surge around your brain, and the electric signals they produce can be picked up by another type of modified oscilloscope called an electroencephalograph (EEG).

The EEG is linked by wires to electrodes that are placed on the patient's head. The machine detects the differences in voltage between various pairs of electrodes and records them as a wavy line on a paper graph or displays them on a computer screen. In a resting adult, for example, the commonest patterns are called alpha waves, which repeat 8 to 12 times per second. Different patterns take over when we are asleep or thinking hard about some problem. EEGs are useful in diagnosing conditions such as epilepsy.

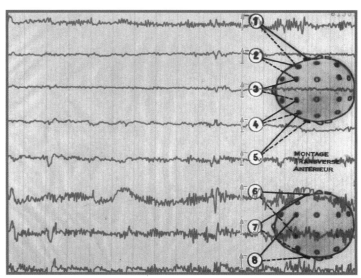

▲ **An electroencephalograph (EEG)** picks up the electrical activity of the brain through electrodes placed on the person's head. The resulting "brain waves" can be recorded as a series of wavy lines on a chart.

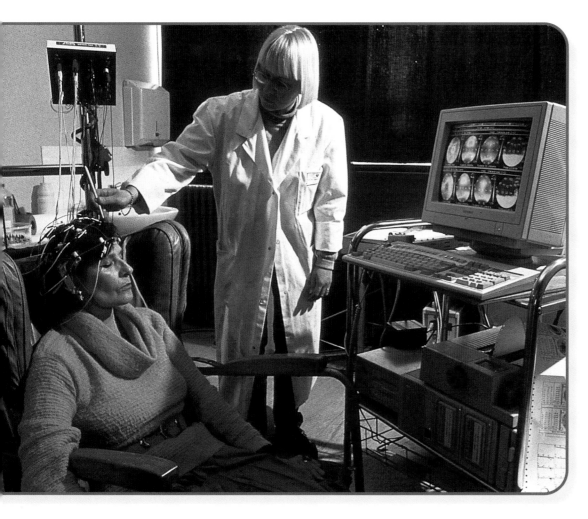

◄ **A computer** can produce a different kind of electro-encephalogram (EEG). Here the screen displays the examination results as a series of color-coded cross-sections of the patient's brain.

Measuring resistance

The resistance of a material is a measure of how well it will let an electric current pass through it. Materials that will easily allow high currents to pass through them have a low resistance and are therefore good conductors. Poor conductors, on the other hand, have a high resistance.

Measuring electrical resistance directly is difficult. Usually, scientists construct a circuit and observe how a particular resistance affects the voltage between certain points in the circuit. One such circuit is pictured on the right.

The meter bridge

This resistance-measuring circuit gets part of its name from the 1-meter length of resistance wire with the sliding contact. The wire also acts as a kind of electrical bridge, as you can see from the circuit diagram.

To use the bridge, a battery is connected across the ends of the wire, which is also connected to thick metal conductors and two pairs of terminals (see the illustration). A known resistance is connected between two of the terminals, and the resistance to be measured—the unknown resistance—is connected across the other two. Finally, a galvanometer is connected from a point between the resistors to a sliding contact on the wire.

The sliding contact is moved until there is no deflection on the galvanometer, when the bridge is said to be balanced. The ratio of the lengths of wire on either side of the sliding contact (shown red and blue) is then the same as the ratio of the values of the resistances. From the wire lengths and the value of the known resistance the value of the unknown resistance can be calculated. The math goes like this:

$$\frac{\text{Unknown resistance}}{\text{Red length}} = \frac{\text{Known resistance}}{\text{Blue length}}$$

Therefore

$$\text{Unknown resistance} = \frac{\text{Known resistance} \times \text{Red length}}{\text{Blue length}}$$

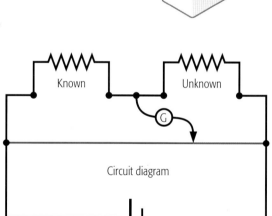

Galvanometer

Unknown resistance

Known resistance

Sliding contact

Resistance wire

Battery

Known   Unknown

G

Circuit diagram

▲ **A meter bridge**, which is used for measuring resistance, gets its name from the 1-meter length of resistance wire along it.

▶ **Line workers** carry instruments to measure the resistance of overhead power lines.

Measuring resistance

▼ **A potential divider** provides a way of varying the output voltage of a battery, which is "divided" by the sliding contact.

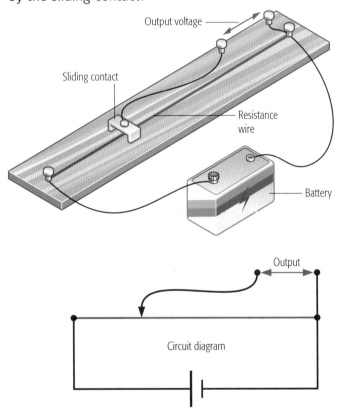

Output voltage

Sliding contact

Resistance wire

Battery

Output

Circuit diagram

Potential divider

This simple circuit uses a 1-meter resistance wire with a sliding contact as a way of varying the output voltage of a battery (see the illustration on the left). The battery is connected between the ends of the resistance wire. There are also two other terminals, one connected to one end of the resistance wire and the other to the sliding contact on the wire.

The sliding contact is moved along the wire to "tap off" the required fraction of the battery's voltage. In the illustration the length of the blue part of the wire is about five-eighths of the total length. The voltage between the terminals (the output voltage) is therefore five-eighths of the voltage of the battery. With a 12-volt battery, for example, the output voltage would be 7.5 volts.

▼ **The brightness of a spotlight** in theaters and outdoor displays is controlled by a variable resistance. Called a dimmer, the resistance controls the voltage to the spotlight—the higher the voltage, the brighter the light.

Making an ohmmeter

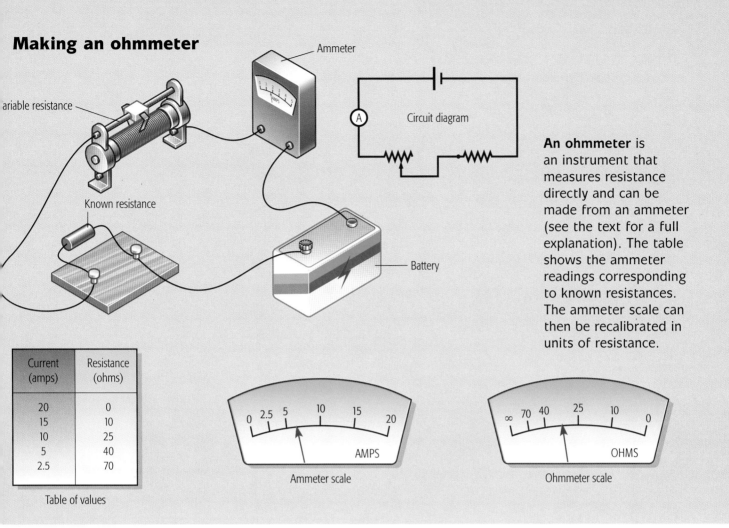

Ammeter

Circuit diagram

An ohmmeter is an instrument that measures resistance directly and can be made from an ammeter (see the text for a full explanation). The table shows the ammeter readings corresponding to known resistances. The ammeter scale can then be recalibrated in units of resistance.

ariable resistance

Known resistance

Battery

Current (amps)	Resistance (ohms)
20	0
15	10
10	25
5	40
2.5	70

Table of values

Ammeter scale

AMPS

Ohmmeter scale

OHMS

Ohmmeter

An ohmmeter is an instrument that measures resistance directly. The illustration above shows how to make one by recalibrating an ammeter (altering its scale). In the circuit are a battery, an ammeter, a variable resistance, and a pair of terminals to which the resistance to be measured will eventually be connected.

To begin with, the two terminals are joined by a length of thick wire so that there is no resistance at all between them. The variable resistance is then adjusted until there is a full-scale deflection on the ammeter, say 20 amps. This is the current that flows around the circuit when there is no resistance at all between the two terminals.

Known resistances of 10, 25, 40, and 70 ohms (say) are then connected between the terminals in turn, and the ammeter readings noted (the table above gives the actual values). The scale on the ammeter can now be changed to these ohm values.

Then, when an unknown resistance is connected between the terminals, its value will be indicated on the scale—the instrument has become an ohmmeter. Notice that the zero on the resistance scale is on the right (not on the left, as on the ammeter scale). Zero on the ammeter scale corresponds to "∞" (infinity) on the ohmmeter scale: If no current can flow, the resistance must be infinitely high.

Electrical and electronics engineers often use an instrument called a test meter or multimeter that can measure current, voltage, or resistance. It is in fact a milliammeter with resistances that can be switched in to vary the amps range or to convert it into a voltmeter (as described on page 20). An internal battery and other resistances allow it to function as an ohmmeter in the way just described. Again, the zero of the ohms scale is at the opposite end from the zeros of the amps and volts scales. One of the commonest uses for

FOR MORE ON MEASURING RESISTANCE SEE *MEASURING CURRENT* **4**:*14*; *MEASURING VOLTAGE* **4**:*20*; *MEASURING ELECTRIC POWER* **4**:*32*

the resistance scale on a multimeter is to test for breaks in a circuit—a continuous circuit has little or no resistance, while a broken circuit has infinite resistance. For instance, a light bulb that has infinite measured resistance must have a broken filament and be no good. Electrical fuses can be tested in a similar way.

Resistance thermometers

The electrical resistance of any material is not constant, but changes with temperature. The hotter a metal gets, the higher is its resistance. For example, the resistance of platinum triples when it is heated through 500°C. This property is used in a device for measuring temperature called a resistance thermometer.

A platinum resistance thermometer has a coil of platinum wire inside a glass tube. Copper wires joined to the platinum are connected to an instrument for measuring resistance, or they can be connected as the "unknown" resistance in a bridge circuit like the one illustrated on page 26. The resistance can be measured very accurately, so a platinum resistance thermometer can detect very small changes in temperature. It can measure temperatures over a large range, from –200 to 1,200°C (about –330 to 2,200°F).

Georg Simon Ohm

The German physicist Georg Ohm was born at Erlangen in 1827, the son of a mechanic. He studied at the university there and, after holding various minor posts, became professor of physics at Munich in 1849. Long before this, in 1827, he had published the famous law that now bears his name, which states that the current flowing through a conductor is proportional to the potential difference across it. Put another way, the voltage divided by the current is a constant, known as the resistance of the conductor. It was 20 years before the importance of Ohm's law was recognized. Ohm died in 1854. The SI unit of resistance is called the ohm in his honor.

◄**Because of its resistance** a length of wire gets hot when electricity flows along it. This effect is used on all scales, from industrial furnaces to the electric oven used to cook a pizza.

►**An electrician's test meter** is also called a multimeter because, by turning the rotary switch, it can act as an ammeter, voltmeter, or ohmmeter (to measure current, voltage, or resistance).

Measuring electric power

The unit of electric power is the watt. It is named after James Watt, the Scottish engineer who invented the steam engine. The watt is rather a small unit, and for many practical purposes the kilowatt (a thousand watts) or even the megawatt (a million watts) is more convenient.

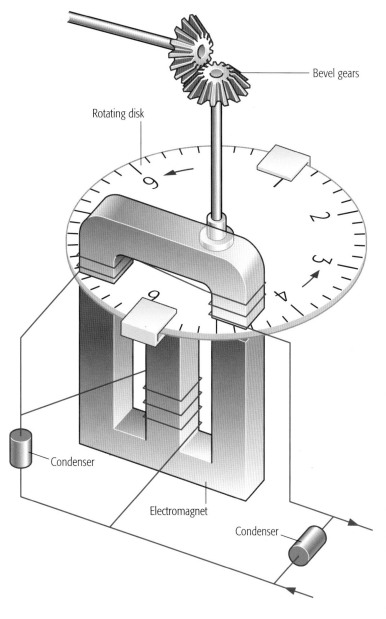

Rotating disk

Bevel gears

Condenser

Electromagnet

Condenser

▲ **A wattmeter** measures electric power consumption. Rotations of the central disk turn gears that work a mechanical or digital counter.

Electricity is a form of energy, and a very adaptable one because it can easily be changed into other forms of energy. In an electric light bulb it is converted into light; in an electric motor it changes into mechanical energy, in an oven it turns into heat; and when an accumulator (storage battery) is charged, electricity is changed into chemical energy. When electricity moves from one point in a circuit to another point that is at a different potential (voltage), there is a change in energy. In other words, work has to be done to move electric charge. The amount of work equals the voltage times the charge, which is the same as the voltage multiplied by the current multiplied by the time for which the current flows.

This work is the same as the energy liberated, as heat, light, or whatever, and is measured in joules (the SI unit of energy). In physics power is the rate of doing work, so electric power is the rate at which energy is liberated. A rate of 1 joule per second is called a watt, and electric power is measured in watts. For anything that uses electricity the power consumed is equal to the voltage multiplied by the current (watts = volts × amps). For large amounts of power the usual unit is the kilowatt (1,000 watts) or the megawatt (1,000,000 watts).

The wattmeter

The technical name for the electric meter in your home is a wattmeter—because it measures watts—and its job is to record how much electric power you use. The meter is connected to the

▲ **An Apollo 12 astronaut** on the surface of the Moon in 1969 setting up equipment that used electricity from low-power solar cells.

power supply, and the current energizes an electromagnet in the instrument. This causes a disk to rotate (it is a type of electric motor). Rotations of the disk turn gears that work a digital counter or dials on a mechanical counter. On the type with dials you can see the edge of the disk turning when an electric appliance is switched on.

Consumption of electrical energy is measured in kilowatt-hours. A rate of 1 kilowatt per hour is called a unit, which is used by the electric company as a basis for calculating your electricity bill. Some home appliances use a lot more electric power than others do. The table on the right lists typical power consumptions. From these figures you can see what a waste of electricity occurs when the hot tap is left running!

APPLIANCE	POWER CONSUMPTION
Light bulb (100 watt)	$\frac{1}{10}$ unit
Heater	1 unit
Icebox	$\frac{1}{24}$ unit
Iron	1 unit
Kettle	2 units
Oven	2 units
Radio	$\frac{1}{10}$ unit
Sewing machine	$\frac{1}{20}$ unit
Shower	1 unit
Television	$\frac{1}{6}$ unit
Toaster	1 unit
Vacuum cleaner	$\frac{1}{4}$ unit
Water heater	3 units

FOR MORE ON MEASURING ELECTRIC POWER SEE *MEASURING CURRENT* 4:14; *MEASURING VOLTAGE* 4:20; *MEASURING RESISTANCE* 4:26

Measuring electric power

Heating effect

When an electric current flows along a wire, some of the electrical energy is converted to heat, and the wire may get warm. The filaments in light bulbs and the elements in electric heaters are made from high-resistance wire, and so get very hot indeed. But ordinary wires, such as the wiring in your home or the cords to various appliances, can get dangerously hot if they carry too much current. Wires that are too thin are the commonest cause of electrical fires.

To guard against this risk, electrical circuits contain a "weak link" in the form of a fuse. It is a piece of thin wire (usually made of a tin–lead alloy) that gets hot and melts and breaks the circuit if too much current flows. There are fuses in the fuse box where the power line enters a home, and the plugs on the cords of some appliances have their own individual fuses.

A 240-volt power circuit is often protected by a 13-amp fuse. A 2-kilowatt heater plugged into the circuit will draw a current of 2,000 ÷ 240 = 8.3 amps. But if you plug two of these heaters into the same circuit, the doubled current of 18.6 amps will "blow" the 13-amp fuse. If the fuse does not blow, the wiring could get dangerously hot and start a fire.

Transmitting power

The actual heat produced by electricity is proportional to the square of the current, so that if the current is doubled, the heat goes up by four times. This heating effect has important consequences for the design of the power lines that carry electricity around the country.

As we have seen, power (watts) equals current (amps) multiplied by voltage (volts). So a given amount of power, say 10 kilowatts (10,000 watts),

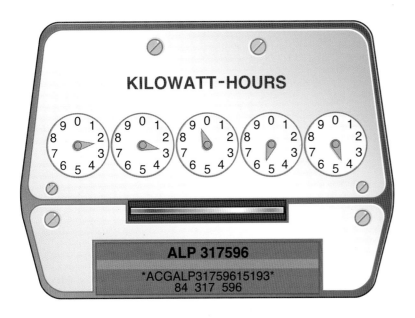

▲ **There are two kinds** of home electric meter, which measure power consumption in kilowatt-hours. In the older, clock-type meter (above) the power is indicated by dials. The more modern digital meter (right) displays numbers to indicate the power. Both of these meters read the same at 22,954 kilowatt-hours.

Thomas Alva Edison

Edison was one of the greatest inventors of his age. By the time he died in 1931, he had more than 1,300 patents to his name. He was born in Milan, Ohio, in 1847 and was taught at home by his mother—he did not do well at school because of his deafness. He moved to Port Huron, Michigan, in 1854, and as a teenager he used to sell newspapers on the railroad there. He grew interested in the telegraph and was soon inventing his own equipment. Edison set up a laboratory at Menlo Park, New Jersey, where he made some of his best-known inventions, including the phonograph and the electric light bulb. He was the first to see the value of electric power and opened the world's first permanent electric power plant in New York in 1882. Ten years later Edison's various companies merged to form the General Electric Company (GE).

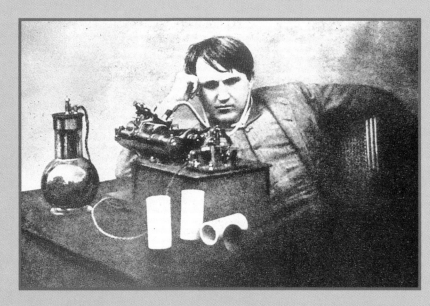

▲ **The inventor Thomas Edison** sits listening to his phonograph. Recordings were made on wax cylinders—some spare ones are visible in front of the machine.

could be supplied either at a low current and a high voltage or at a high current and a low voltage. In our example the power line could carry 1,000 volts at 10 amps or 100 volts at 100 amps. But the second combination would produce a hundred times more heat than the first one and result in a tremendous waste of energy—hot power lines merely heat the air or the ground around them. For this reason commercial electricity distribution uses high voltage and low current.

Although electricity is often generated at high voltages at the power plant, which is ready for

high-voltage distribution, much lower voltages are needed by industry and homes. It is the job of transformers in local substations to change the high voltages to lower ones (see the picture on page 8). Transformers are important electrical devices. As well as the large ones used in the distribution of electric power, many items of electronic equipment also include them. Radios, TVs, stereo systems, and personal computers all need transformers to change the supply voltage (240 volts) into the very low voltages needed for the circuits in such equipment.

Measuring magnetic fields

A magnetic field is the region around a magnet in which its effect can be felt. This definition applies to small magnets, such as those that hold the door shut on the icebox, and to the powerful magnetic fields that surround the Sun, the Earth, and some other planets.

Compasses are the oldest science tools that make use of a magnetic field—the Earth's magnetic field. The Earth behaves as if it had a giant bar magnet along its axis between the North and South Poles. The field curves around from one pole to the other. In a compass the pointer (called the needle) is itself a small magnet, pivoted at its center so that it can swing from side to side. When it points to the north, it is in fact indicating the direction of the Earth's magnetic field.

▲ **A maglev train** makes practical use of magnetic fields to hold it clear of the rails and propel it along smoothly and quietly.

Magnetometer

A compass is not the only scientific instrument that uses magnets. We have already seen (page 18) that a tangent galvanometer detects electric currents by using a small pivoted magnet. Fairly similar in this respect is a magnetometer, which, as its name suggests, is an instrument for measuring the strength of a magnetic field.

The instrument consists of a small pivoted magnet attached at right angles to a long, nonmagnetic pointer. It is turned until the magnet lines up with the Earth's magnetic field, and the pointer indicates zero (illustration A). When another magnetic field, such as that of a bar magnet, is nearby, one end of the magnet is attracted to it (because opposite magnetic poles, a north and a south, attract each other). This makes the pointer move around the scale (illustration B). The angle indicated by the pointer is a measure of the strength of the external field (the actual field strength is proportional to the tangent of the indicated angle).

Just as opposite magnetic poles attract each other, similar ones—two north poles or two south poles—repel each other. This effect is put to good use in a maglev train. Maglev is short for "magnetic levitation" (lifting), and similar electromagnets inside the train lift the vehicle clear of the rails. Electromagnetic effects like this are used in a wide range of devices, including electric bells, relays, loudspeakers, generators, and electric motors.

S ▬▬▬▬ N

Magnet

Magnetic detectors

The presence of nearby metal objects can alter a magnetic field. That is what makes metal detectors work—you may have seen them in airports or stores. The apparatus consists of a tall archway the size of a door. Large coils of wire in the sides of the arch carry an electric current, which creates a magnetic field.

▼ **A magnetometer** consists of a small pivoted magnet attached at right angles to a long pointer that indicates the direction of the magnetic field (A) of the Earth and (B) of a nearby magnet.

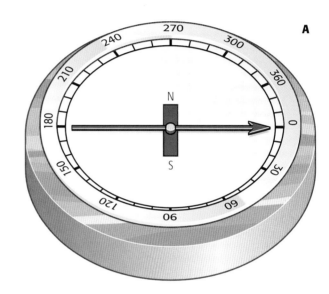

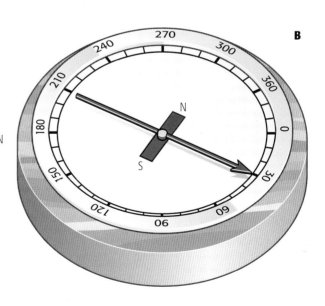

Anyone who walks through the arch carrying a metal object, such as a handgun, will disturb the magnetic field. The change in the field is detected and causes an alarm to sound. A bunch of keys or a metal watchband will also

FOR MORE ON MEASURING MAGNETIC FIELDS SEE *ELECTRON MICROSCOPES* **6**:20; *SPECTROSCOPES* **8**:20; *ROCKS AND MINERALS* **9**:14

Measuring magnetic fields

▲ **A scientist** uses a metal detector to search the desert for fragments of fallen meteorites—objects from outer space that often contain metals such as nickel and iron.

set off the alarm, which is why people are asked to remove such objects before they walk through the detector.

People searching for buried objects use a different type of metal detector. It consists of a coil of wire carrying an electric current mounted inside a disk at the end of a long handle. Batteries in the handle supply the electricity. The coil produces a magnetic field that is distorted by any metal objects in the ground. A microchip in the handle detects the tiny changes in current produced by the field distortion and causes the detector to make a buzzing or bleeping sound. Scientific versions of the instrument give an audible warning via earphones, as in the picture on the left.

More sensitive magnetometers are used in prospecting for underground minerals. To cover a large area quickly, the instrument is mounted in an aerodynamic "torpedo" and towed behind a light airplane or helicopter. It works by detecting distortions in the Earth's magnetic field (which is usually fairly regular) caused by mineral deposits. Minerals nearer the surface have a greater effect than those buried deep underground. The data from the magnetometer passes along the cable to an onboard computer in the towing aircraft.

► **A metal detector** works by creating a magnetic field and responding to changes in the field caused by the presence of a metallic object, such as this buried horseshoe.

► **Minerals under the ground** cause small changes in the local magnetic field. A magnetometer towed behind a small airplane or helicopter can detect these changes and pinpoint areas that might contain useful minerals.

Glossary

Any of the words in SMALL CAPITAL LETTERS can be looked up in this Glossary.

accumulator Also called a *storage battery*, a type of secondary cell, which is an electric cell (battery) that, when it runs down, can be recharged by connecting it to an external supply of current.

ammeter An instrument for measuring ELECTRIC CURRENT.

ampere (symbol **A**) The SI unit of ELECTRIC CURRENT, often abbreviated to "amp," equal to a rate of current flow of 1 COULOMB per second.

anode The positive ELECTRODE in an electric cell (battery) or ELECTROLYSIS cell. It is the electrode through which ELECTRONS enter the cell.

battery See ACCUMULATOR.

capacitance The ratio of the stored electric charge on an electrical device or other object to the POTENTIAL DIFFERENCE applied to it.

capacitor Also called a *condenser*, a device (circuit component) that stores electric charge.

cathode The negative ELECTRODE in an electric cell (battery) or ELECTROLYSIS cell. It is the electrode by which ELECTRONS leave the cell.

cathode-ray tube A vacuum tube in which electric or magnetic fields deflect a stream of ELECTRONS from an ELECTRON GUN and make it "draw" lines on the PHOSPHOR coating inside the flat front face of the tube.

compass A device with a pivoted magnetized needle that always swings around to point to MAGNETIC NORTH.

condenser Another name for a CAPACITOR.

conductor A material or circuit component that ELECTRIC CURRENT can flow through.

coulomb (symbol **C**) The SI unit of electric charge, equal to the charge carried by 6.24 billion billion ELECTRONS.

coulometer Another name for a VOLTAMETER.

dosimeter A small device for detecting ionizing radiation that records the amount of radiation exposure and is worn like a badge by people who work with x-rays and in the nuclear industry.

electric current A flow of ELECTRONS through a CONDUCTOR.

electric potential The amount of energy a positive unit charge would gain if it were brought from infinity to a point in an electric field where it is being measured. It is measured in VOLTS.

electrocardiograph (ECG) An instrument for making a record (an *electrocardiogram*) of the electrical activity of the heart.

electrode A metal plate or carbon rod through which ELECTRONS enter or leave an electric cell (battery), ELECTROLYSIS cell, or vacuum tube.

electroencephalograph (EEG) An instrument for making a record (an *electroencephalogram*) of the electrical activity of the brain.

electrolysis The chemical decomposition of an ELECTROLYTE when an electric current passes through it.

electrolyte A liquid (either a solution or a molten substance) that conducts electricity, as in ELECTROLYSIS.

electrometer An instrument for measuring differences in electric charge or voltage.

electron A SUBATOMIC PARTICLE that has a negative charge; it is the basic unit of electricity. Electrons surround the NUCLEUS of an atom.

electron gun A device for producing a stream of ELECTRONS in, for example, a CATHODE-RAY TUBE, an x-ray tube, or an electron microscope.

electroscope An instrument for detecting electric charge.

farad (symbol **F**) The SI unit of CAPACITANCE, equal to the capacitance of a capacitor with a charge of 1 COULOMB and a POTENTIAL DIFFERENCE of 1 VOLT between its plates.

free electron In the atoms of a CONDUCTOR (such as a metal), an outer ELECTRON that is free to move and carry an ELECTRIC CURRENT.

fuse A circuit component that melts and breaks the circuit (to protect it) if too much current flows.

galvanometer An instrument that detects and measures very small ELECTRIC CURRENTS.

henry (symbol **H**) The SI unit of INDUCTANCE.

hot-wire ammeter A type of AMMETER that relies for its action on the expansion of a piece of wire that is heated by the electric current flowing in it.

inductance The generation of a current in a circuit by changes in a nearby magnetic field. It is measured in HENRYS.

insulator Also called a *nonconductor*, a substance that is a poor conductor of ELECTRIC CURRENT (because it lacks FREE ELECTRONS).

kilowatt-hour (symbol **kWh**) Also called a *unit*, the consumption of 1,000 watts (1 kilowatt) of electric POWER for 1 hour.

kilovolt (symbol **kV**) A unit of potential difference (voltage) equal to 1,000 VOLTS.

kilowatt (symbol **kW**) A unit of electric POWER equal to 1,000 WATTS.

maglev Short for **mag**netic **lev**itation, a system of railroad propulsion that uses magnetic fields to lift a train clear of the rails and propel it along.

magnetic field The region around a magnet (or electromagnet) in which its influence can be felt.

magnetic north The direction indicated by the needle of a compass. It is not exactly the same as true north and changes position slightly over the years.

magnetometer An instrument for measuring the strength (and sometimes the direction) of a MAGNETIC FIELD.

megawatt (symbol **MW**) A unit of electric POWER equal to 1,000 WATTS.

metal detector A device that detects the presence of a metal object by its effect on a strong magnetic field.

meter bridge An electrical apparatus consisting of a meter-long piece of high-resistance wire with a sliding contact, used for measuring RESISTANCES.

microfarad (symbol μF) A unit of CAPACITANCE equal to one-millionth of a FARAD.

milliammeter A type of AMMETER for measuring very small ELECTRIC CURRENTS.

milliampere (symbol **mA**) A unit of ELECTRIC CURRENT equal to one-thousandth of an AMPERE.

millivolt (symbol **mV**) A unit of POTENTIAL DIFFERENCE (voltage) equal to one-thousandth of a VOLT.

milliwatt (symbol **mW**) A unit of electric POWER equal to one-thousandth of a WATT.

moving-coil ammeter A type of AMMETER in which the current to be measured causes a coil to rotate in a magnetic field. A pointer connected to the coil indicates the current on a scale.

moving-coil voltmeter A type of VOLTMETER that consists of a MOVING-COIL AMMETER with a high resistance in series with it.

moving-iron ammeter A type of AMMETER that relies for its action on the movement of a piece of iron that is magnetized by the current it is measuring.

multimeter An electrical measuring instrument that is a combination of an AMMETER, a VOLTMETER, and an OHMMETER.

nonconductor Another term for an INSULATOR.

ohm (symbol Ω) The SI unit of RESISTANCE.

ohmmeter An instrument for measuring electrical RESISTANCE.

oscilloscope An electrical instrument that displays a varying electrical signal as a wavy line on the face of a CATHODE-RAY TUBE.

phosphor A chemical that fluoresces (gives off light) when struck by streams of electrons. Phosphors are used on the inside of the face of a CATHODE-RAY TUBE and in fluorescent lamps.

picofarad (symbol **pF**) A unit of capacitance equal to a million-millionth of a FARAD.

platinum resistance thermometer A type of thermometer that relies on the fact that the electrical resistance of a platinum wire changes with changes in temperature. A measure of the resistance is therefore a measure of the temperature.

potential difference Also called *voltage*, the difference in ELECTRIC POTENTIAL between two points (in a circuit). The higher the potential difference, the greater the force tending to move charges (such as the electrons that make up a current) between the two points.

potential divider Also called a *potentiometer*, an electrical apparatus consisting of a length of resistance wire that can be tapped by a sliding contact to obtain a known fraction of the total voltage across the ends of the wire.

potentiometer Another term for a POTENTIAL DIVIDER.

power The rate at which electrical energy is produced or consumed. Its SI unit is the WATT.

resistance A measure of how much a material or component resists the passage of an ELECTRIC CURRENT through it. The higher the resistance, the less current will pass for a given potential difference (voltage) across it. It is measured in ohms.

resistance thermometer See PLATINUM RESISTANCE THERMOMETER.

resistor An electrical circuit or circuit component that has RESISTANCE.

subatomic particle Any of the particles that make up atoms, including ELECTRONS, NEUTRONS, and PROTONS.

storage battery An alternative term for an ACCUMULATOR.

tangent galvanometer An electrical instrument that detects and measures very small electric currents by their effects on a pivoted magnet.

transformer An electric device that converts an alternating voltage (AC) into one that is higher (a *step-up* transformer) or lower (a *step-down* transformer).

unit Another term for KILOWATT-HOUR.

volt (symbol **V**) The SI unit of POTENTIAL DIFFERENCE (voltage).

voltage Another term for POTENTIAL DIFFERENCE.

voltameter An apparatus for measuring quantity of electric charge in terms of the mass of a metal deposited (on the CATHODE) during ELECTROLYSIS.

voltmeter An instrument for measuring POTENTIAL DIFFERENCE (voltage).

watt (symbol **W**) A rate of energy production or consumption equal to 1 joule per second. (The joule is the SI unit of energy.) In an electric circuit it is equal to the ELECTRIC CURRENT multiplied by the POTENTIAL DIFFERENCE.

wattmeter An instrument for measuring electric POWER consumption (usually in KILOWATT-HOURS, or units).

Set Index

Set Index

Fungi **9**:23, *32*
furlong **1**:14, 16
fuses **4**:30

G

galaxies **3**:7; **6**:38; **9**:34, *34*, 36, *36*, 37, *37*
Galileo Galilei **2**:19, 28, **3**:*24*, **5**:22, *24*
see also telescopes
gallium **9**:12
gallium arsenide **6**:16
galvanometer **4**:18, 21, *22*, 26, *26*
tangent **4**:18, *18*, 37
gamma rays
Earth's atmosphere **6**:10, *11*
frequency **6**:9
production **6**:27, 30
radioactivity **6**:6–7, 27
Sun **6**:10, *11*
wavelength **6**:6–7
gas chromatography **8**:*14–15*
Gascoigne, William **1**:12
gases
conductivity **3**:21
density **3**:8
pressure **3**:20–25
gastropods **9**:28
gauges **1**:*9*, 11, 12, *12*
strain gauge **2**:25
see also pressure gauges
gear wheels **2**:20–21, *21*
gears, car **2**:32–33
Gell-Mann, Murray **3**:36
genera **9**:26
General Electric Company **4**:*35*
generators **3**:14; **4**:9
genetic fingerprinting **8**:28, 28–31
genetics **8**:8
genus **9**:18
geology
faults **7**:*32*, *33*, 34
Grand Canyon **2**:*34*
microscopes **5**:21
see also earthquakes; minerals; rocks
geometry **1**:21
germanium **9**:12
geyser **3**:*22*
gigawatt **4**:9
ginkgo **9**:22
Glaser, Donald **3**:37
Global Positioning System **1**:39
gluons **3**:*36*, 37
Gnetophyta **9**:22–23
gold **3**:8
Graham, George **1**:12
Gram, Hans **9**:33
grams **3**:6
Gram's stain **9**:33
Grand Canyon **2**:*34*
granite **2**:*34*; **9**:17
Grant, George **2**:21

gravity **3**:10
Earth **3**:15, 18
effects **3**:12–13
mass **3**:6, 11
Moon **3**:15, 18
Newton's law **3**:*10*, 11, *13*
rocket launch **3**:*14*
Solar System **3**:15
weight **3**:6–7
Greeks
balance **3**:18
Ctesibius **3**:26
sundial **2**:16–17
water clock **2**:14, *14*
Green, Andy **2**:32
Greenwich Mean Time **1**:35, **2**:26, *30*, *39*
Gregory, James **5**:29
Gregory XIII, pope **2**:10
Grimaldi, Francesco **5**:11
guitar **7**:23
Gunter's chain **1**:14
Guo Shoujing **2**:17
gymnosperms **9**:21–22

H

hair, static electricity **4**:10, *10*
hairsprings **2**:20; **4**:15, *15*, 17, *17*
Hale, George Ellery **5**:29
halite **3**:*33*
hardness test **9**:16–17
Harrison, John **2**:28, 29
Hawaii, Mauna Kea **5**:*30*, 32; **6**:19
headset **7**:*10*, 22
hearing aids **7**:11
heart **3**:30; **4**:23
heartbeats **4**:23–24
heat
infrared radiation **6**:8
heat rays **6**:8, 10, 14
heating effect **4**:14, 16, 34–35; **8**:23
heights **1**:*6–7*, *7*, 20–23, 25
helium **3**:*36*; **8**:*14*, 15
Henlein, Peter **2**:20
Henry, Joseph **4**:8
Henry, Sir Edward **9**:9
henrys **4**:8
herbivores **9**:26
Herschel, Caroline **6**:*15*
Herschel, Sir William **5**:29; **6**:15, *15*
hertz **5**:8; **6**:9; **7**:6, 8, 21, 24
Hertz, Heinrich **7**:8
Hertzsprung, Ejnar **9**:35
Hertzsprung-Russell diagram **9**:35, *35*, 36
hesperidium **9**:20, *20*
hikers **1**:*35*
Hillier, James **6**:25
Hipp, Matthäus **2**:21
Hooke, Robert **2**:20; **3**:*24*; **5**:16
horizon glass **1**:22

horoscopes **2**:11
horse **9**:*27*
horse races **1**:16
horsetails **9**:22
Horton, Joseph W. **2**:24
hourglass **2**:12, *12*, *13*
hours **2**:9
Hubble, Edwin **9**:*36*, 37, *39*
Hughes, David Edward **7**:14, *14*
humans
blood **9**:30–31
evolution **9**:9
hearing **7**:32
teeth **9**:*27*, 28
hundredweights **3**:6
hurricanes **8**:*36*, 38
Huygens, Christiaan **2**:19, 20
hydraulics **3**:26, 30
hydrochloric acid **8**:*12*, 13, 16–18
hydrogen **3**:8; **8**:15; **9**:10
hydrogen sulfide **8**:*12*, 13
hydrophone **7**:28
hydroxides **8**:*12*, 13, 19
hypodermic needle **8**:*21*

I

IBM Research Laboratory **6**:25
image intensifier **6**:*16*
Improved TOS satellites **6**:18
inches **1**:*6–7*, 13, 29
index glass **1**:22
India **7**:*37*
inductance **4**:8
inertia **3**:6
Infrared Astronomical Satellite (IRAS) **6**:*18*, 19
infrared radiation **6**:*11*, 14–15
detectors **5**:36; **6**:8, 15–16, *18*, 19
heat **6**:8, 10
production of **6**:16
satellites **6**:18–19
spectrum **6**:*17*, 18; **8**:*23*
submillimeter **6**:18
infrasound
animals **7**:*38*, 39
earthquakes **7**:32, 35, *36*, 36–37
spiders **7**:38
insects **7**:26; **9**:7, *9*, 24, 25, 29
insulators **4**:7
interference **5**:10, 12, *12*
interferometer **6**:*38*, 39
International Bureau of Weights and Measures **3**:6
International Date Line **1**:36; **2**:*38*
International Ultraviolet Explorer **6**:13
invertebrates **9**:25, 28

iodine **9**:11, 12, 13
ionic bonds **3**:*32*, 32–33, *33*
ions
bonds **3**:33, *33*
charged **4**:13; **8**:15, 8:26–27
electrolysis **4**:*18*
electrons **8**:26–27
helium **3**:36
metals **4**:13; **8**:*12*
iron hydroxide **8**:19, *19*
iron oxide **8**:19, *19*
isobars **1**:37
isotopes **2**:34, **3**:38–39, *39*
infrared lasers **6**:16
radioactivity **2**:34; **3**:39
radiocarbon dating **2**:36–37

J

Jansky, Karl **6**:36
Japanese Subaru Telescope **6**:19
JNWP weather forecasting **8**:39
joules **4**:32
Jupiter **3**:15; **5**:22; **9**:38, *38–39*

K

keyboards **7**:*23*
kidney stones **7**:25
kilogram per cubic meter **3**:8
kilograms **3**:6
kilometers **1**:*28*, *24*, 6, 13
kiloparsecs **1**:19
kilopascals **3**:8
kilovolts **4**:8
kilowatt-hours **4**:33, *34*
kilowatts **4**:9, 32
kingdoms **9**:*32*
Kirchhoff, Gustav **8**:22
Kitt Peak National Observatory **5**:32

L

laboratory **9**:*11*
Landsat **1**:*37*; **6**:*18*; **1**:*36*
Landsteiner, Karl **9**:30, 31, *31*
Langevin, Paul **7**:29
Langley, Samuel Pierpont **6**:15–16
lanthanides **9**:*12*
Large Electron Positron Collider **3**:37–38, *38*
lasers **1**:14, *15*; **6**:16; **7**:19, *19*
latitude **1**:22, 34–35, 36; **2**:17, *28*
lawrencium **9**:*12*
lead **8**:21
lead sulfide **6**:16
leadsman **7**:28
leap seconds **2**:39
leap years **2**:8
leaves **9**:19, *19*
LEDs **6**:16

legumes **9**:20, *20*
length **1**:*6–7*, 13
lenses **5**:14–15, 27, *27*
cameras **5**:34–35
concave **5**:*14*, 14–15, 26
condenser **5**:38, *38*
convex **5**:*14*, 14–15, *15*, 22, 25
diverging **5**:*14*, 15, 26
electron **6**:24
erector **5**:27, *27*
eyepiece **5**:16, *17*, *18*, 20, 23, 24, 25–26
fish-eye **5**:36
magnetic **6**:21
objective **5**:16, *17*, *18*, 19, 23–24, 25–26
telephoto **5**:36, *37*
ultrawide-angle **5**:36
Van Leeuwenhoek **5**:*21*
wide-angle **5**:35–36
see also microscopes; optical instruments; telescopes
leopard **1**:*10*; **9**:*24*
Libby, Willard **2**:*36*, 37
library classification **9**:8–9
light
beam **5**:8, 10, 11
colors **5**:8–9, *8–9*
diffraction **3**:35, *35*; **5**:10, 11, *11*
focus **5**:14–15
haze **6**:14
infrared **6**:18–19
interference **5**:10, 12, *12*
lenses **5**:14–15
microscopes **5**:*17*, 18–19, *20*
radiation **5**:7–8; **6**:6
rays **5**:*14*, 15, 28
refraction **5**:8, 9, 10, 14
speed **5**:8; **6**:9
ultraviolet **6**:6, 7, 13; **8**:31
waves **5**:6–9; **8**:24
white **5**:*6*, 7, *9*, 10, 12, 25
light radar (LIDAR) **6**:16
light years **1**:13, 19
lightning **4**:*13*
limestone **2**:*34*; **9**:17
line spectra **8**:*22*, 24
line workers **4**:27
linear accelerators **3**:37
Linnaeus, Carolus **9**:18, 23, *23*
Linné, Karl von **9**:23
lion **1**:*10*; **9**:26, *27*
Lippershey, Hans **5**:22, 26
liquid hydrogen **3**:37
liquids **3**:9; **8**:24
lithium **8**:22
location **1**:38, *38*, 39
longitude **1**:34–35, 36; **2**:26–28, 38–39
loudness **7**:6, 7, *7*
loudspeakers **7**:20–23, *21*, *22*

Set Index

Further reading/websites and picture credits

Further Reading

Atoms and Molecules by Philip Roxbee-Cox; E D C Publications, 1992.

Electricity and Magnetism (Smart Science) by Robert Sneddon; Heinemann, 1999.

Electronic Communication (Hello Out There) by Chris Oxlade; Franklin Watts, 1998.

Energy (Science Concepts) by Alvin Silverstein et al.; Twenty First Century, 1998.

A Handbook to the Universe: Explanations of Matter, Energy, Space, and Time for Beginning Scientific Thinkers by Richard Paul; Chicago Review Press, 1993.

Heat (How Things Work Series) by Andrew Dunn; Thomson Learning, 1992.

How Things Work: The Physics of Everyday Life by Louis A. Bloomfield; John Wiley & Sons, 2001.

Introduction to Light: The Physics of Light, Vision and Color by Gary Waldman; Dover Publications, 2002.

Light and Optics (Science) by Allan B. Cobb; Rosen Publishing Group, 2000.

Electricity and Magnetism (Fascinating Science Projects) by Bobbi Searle; Copper Beech Books, 2002.

Basic Physics: A Self-Teaching Guide by Karl F. Kuhm; John Wiley & Sons, 1996.

Eyewitness Visual Dictionaries: Physics by Jack Challoner; DK Publishing, 1995.

Makers of Science by Michael Allaby and Derek Gjertsen; Oxford University Press, 2002.

Physics Matters by John O.E. Clark et al.; Grolier Educational, 2001.

Science and Technology by Lisa Watts; E D C/Usborne, 1995.

Sound (Make It Work! Science) by Wendy Baker, John Barnes (Illustrator); Two-Can Publishing, 2000.

Websites

Astronomy questions and answers — http://www.allexperts.com/getExpert.asp?Category=1360

Blood classification — http://sln.fi.edu/biosci/blood/types.html

Chemical elements — http://www.chemicalelements.com

Using and handling data — http://www.mathsisfun.com/data.html

Diffraction grating — http://hyperphysics.phy-astr.gsu.edu/hbase/phyopt/grating.html

How things work — http://rabi.phys.virginia.edu/HTW/

Pressure — http://ldaps.ivv.nasa.gov/Physics/pressure.html

About rainbows — http://unidata.ucar.edu/staff/blynds/rnbow.html

Story of the Richter Scale — http://www.dkonline.com/science/private/earthquest/contents/hall2.html

The rock cycle — http://www.schoolchem.com/rk1.htm

Views of the solar system — http://www.solarviews.com/eng/homepage.htm

The physics of sound — http://www.glenbrook.k12.il.us/gbssci/phys/Class/sound/u11l2c.html

A definition of mass spectrometry — http://www.sciex.com/products/aboutmass.htm

Walk through time. The evolution of time measurement — http://physics.nist.gov/GenInt/Time/time.html

How does ultrasound work? — http://www.imaginiscorp.com/ultrasound/index.asp?mode=1

X-ray astronomy — http://www.xray.mpe.mpg.de/

Picture Credits

Abbreviation: SPL Science Photo Library

8, 9 U.S. Department of Energy/SPL; **10** Peter Menzel/SPL; **13** Gordon Garradd/SPL; **14** Martin Bond/SPL; **17** Bettmann/Corbis; **18** Sam Ogden/SPL; **19** Simon Fraser/Northumbria Circuits/SPL; **21** Hank Morgan/SPL; **23** Maximilian Stock Ltd/SPL; **25t** BSIP Vem/SPL; **25b** Catherine Pouedras/SPL; **27** John Howard/SPL; **28** Reuters Newmedia Inc/Corbis; **30** Robert Holmes/Corbis; **31** John O.E. Clark; **33** NASA/SPL; **35** SPL; **36** Martin Bond/SPL; **38** David Parker/SPL.